Sawsan El-Sayed
Najla Al-meataq
Kholood Al-faleh and Norah Al-salh

BCK-Algebras

Sawsan El-Sayed
Najla Al-meataq
Kholood Al-faleh and Norah Al-salh

BCK-Algebras

Noor Publishing

Imprint

Cover image: www.ingimage.com

Publisher:
Noor Publishing
is a trademark of
International Book Market Service Ltd., member of OmniScriptum Publishing Group
17 Meldrum Street, Beau Bassin 71504, Mauritius
Printed at: see last page
ISBN: 978-620-2-79336-0

Najla Al-meataq, Norah Al-salh, KholoodAl-faleh,

BCK-algebras

By

Najla Al-meataq, Norah Al-salh, KholoodAl-faleh

DEDICATED TO

MY FAMILY

And especially to

MY FATHER and **MY LOVING MOTHER**

ACKNOWLEDGEMENT

First and foremost, I would like to thank the **Dr. Sawsan El-Sayed**, for her support, outstanding guidance and encouragement throughout my project. I would like to thank our family, especially my mother, for their encouragement, patience, and assistance over the years. We are forever indebted to our parents, who have always kept me in their prayers,

I would like to express my sincere gratitude to Dr. Maram, the coordinator of female section in the mathematics department at the College of Science in Zulfi, Majmaah University for her cooperation. Moreover, I would like to thank all teaching staff in the mathematics department, for their continuous advice and consultations during my sessions. Furthermore, I would like to thank the arbitrators. I would like to thank my colleagues and everyone who has stood by me, practically or spiritually.

Najla Al-meataq,

Narah Al-salh,

KholoodAl-faleh

2021-1442

CONTENTS

Abstract

BCK-algebras have been studied by many authors and they have been applied to many branches of mathematics. In this paper, we introduce the most prominent topics of BCK-algebras and study the relation between the different types of BCK-algebras. In addition, we present the nation of ideals of BCK-algebras and study the relation between the different types of ideals. Finally, we introduce the notion of quotient BCK-algebras.

Keywords

BCK - algebras, subalgebras, bounded BCK-algebra, commutative BCK- algebra, positive-implicative BCK-algebra, implicative BCK-algebra, Ideals of BCK- algebras, Prime ideals, Maximal ideals, positive implicative ideals, irreducible ideals, quotient BCK-algebras.

1. Introduction

In 1966, Imai and Iséki introduced the notion of a *BCK-algebra*. This notion originated from two different ways: (1) set theory, and (2) classical and non-classical propositional calculi. The BCK-operation $*$ is an analogue of the set theoretical difference. Today BCK-algebras have been studied by many authors and they have been applied to many branches of mathematics, such as group theory, functional analysis, probability theory, topology, fuzzy set theory, and so on [1]

In this paper, we introduce the most prominent topics of BCK- algebras such as: subalgebras, bounded BCK- algebra, commutative BCK-algebra, positive-implicative BCK-algebra. In section 1, we introduced the previous studies about BCK- algebra. In section 2, we study the properties of BCK-algebras. In Section 3, we study the subalgebras. In Section 4, we study the bounded BCK- algebra. In Section 5, we study the commutative BCK-algebra. In Section 6, we study the Positive-implicative BCK-algebra. In addition, we introduce the notion of ideals of BCK-algebras and study the different types of ideals. Finally, we present the concept of quotient algebras and its properties.

2. BCK algebras

Definition 2.1 [17]

Let X be a set with a binary operation "$*$" and a constant "0". Then, $(X,*,0)$ is called a BCK-algebra if it satisfies the following conditions: for all $x, y, z \in X$

(BCI – 1) $((x * y) * (x * z)) * (z * y) = 0,$

(BCI – 2) $(x * (x * y)) * y = 0,$

(BCI – 3) $x * x = 0,$

(BCI – 4) $x * y = 0 \; and \; y * x = 0 \; imply \; x = y,$

(BCK – 5) $0 * x = 0.$

Definition 2.2 [17]

We define a binary relation "≤" on X by $x \leq y$ if and only if $x * y = 0$. Then, $(X,*,0)$ is a BCK-algebra if and only if it satisfies the following conditions:

(BCI – 1`) $((x * y) * (x * z)) \leq (z * y),$

(BCI – 2`) $(x * (x * y)) \leq y,$

(BCI – 3`) $x \leq x,$

(BCI – 4`) $x \leq y$ and $y \leq x$ imply $x = y,$

(BCK – 5`) $0 \leq x.$

Example 2.3 [18]

Let $Z = \{0, a, b, c\}$ and define a binary operation $*$ on Z by the following table:

$*$	0	a	b	c
0	0	0	0	0
a	a	0	a	a
b	b	b	0	b
c	c	c	c	0

Then, Z is a BCK-algebra.

Example 2.4 [10]

Let $X = \{0,1,2,3,\dots\}$

$$x * y = \begin{cases} 0 & if \quad x < y \\ 1 & if \; x > y \end{cases}$$

$(X ,* ,0)$ $is\ not\ a\ BCK - algebra$ as $(BCI - 2`)\ is\ not\ satisfied$

$if\ x = 2, \qquad y = 0$

$$[x * (x * y)] * y = [2 * (2 * 0)] * 0$$
$$= (2 * 1) * 0 = 1 * 0 = 1 \neq 0$$

Theorem 2.5 [10]

Let $(X ,* ,0)$ be a BCK algebra. Then, for all x , y , z in X, the following properties are satisfied:

1- $x \leq y\ implies\ z * y \leq z * x$
2- $x \leq y , y \leq z\ implies\ x \leq z$
3- $(x * y) * z = (x * z) * y$
4- $x * y \leq z\ implies\ x * z \leq y$
5- $(x * z) * (y * z) \leq x * y$
6- $x \leq y\ implies\ x * z \leq y * z$
7- $(x * y) \leq x$
8- $x * 0 = x$

Proof:

1) From BCK 1' $(x * y) * (x * z) \leq z * y$
By replacing x and z, $(z * y) * (z * x) \leq x * y$
Since $x \leq y$ then $x * y = 0 \rightarrow (z * y) * (z * x) \leq 0$
By BCK 5' $\rightarrow$ $0 \leq (z * y) * (z * x)$
By BCK 4' $\rightarrow$ $(z * y) * (z * x) = 0$
$\rightarrow z * y \leq z * x$

2) Since $y \leq z\ \ by\ (1)\ x * z \leq x * y$
Since $x \leq y\ then\ x * y = 0 \rightarrow x * z \leq 0$

By BCK 5' $0 \leq x * z$
By BCK 4' → $x * z = 0 \rightarrow x \leq z$

3) From BCK 2' $x * (x * z) \leq z$
By (1) $(x * y) * z \leq (x * y) * [x * (x * z)]$
By BCK 1 $(x * y) * (x * z) \leq (z * y)$
$(x * y) * z \leq (x * y) * [x * (x * z)] \leq (x * z) * y$
$(x * y) * z \leq (x * z) * y$ (i)
By replacing y and z in (i) $(x * z) * y \leq (x * y) * z$ (ii)
By BCK 4' $(x * y) * z = (x * z) * y$

4) Let $x * y \leq z, \rightarrow (x * y) * z = 0.$ From(3)

$(x * z) * y = 0 \rightarrow x * z \leq y$

5) From BCK 1', $(x * y) * (x * z) \leq z * y$

from (4), $(x * y) * (z * y) \leq (x * z)$

interchanging y and z: $(x * z) * (y * z) \leq (x * y)$

6) From (4), $(x * y) * x = (x * x) * y$

From BCK 3, $(x * y) * x = (x * x) * y = 0 * y.$

From BCK 5, $(x * y) * x = (x * x) * y = 0 * y = 0$

$(x * y) * x = 0 \rightarrow x * y \leq x$

7) From BCK 2': $x * (x * y) \leq y$

Put $y = 0$: $x * (x * 0) \leq 0 \rightarrow x \leq x * 0$ ... (i)

By (6) $x * y \leq x \rightarrow x * 0 \leq x$ ... (ii)

From (i) and (ii) $x * 0 = x.$

8) BCK 1', $(x * y) * (x * z) \leq z * y$

By replacing y and z $(x * z) * (x * y) \leq y * z$

Since $x \leq y$ then $x * y = 0 \rightarrow (x * z) * 0 \leq y * z$

Then from (7), $(x * z) * 0 = (x * z) \rightarrow x * z \leq y * z$.

Definition 2.6 [10]

Let $(X,* ,0)$ be a BCK- algebra. For any $x , y \; in \; X$, we define

$$x \wedge y = y * (y * x).$$

Obviously, $x \wedge y$ is a Lower bound of $x \; and \; y$, $and \; x \wedge x = x * (x * x) = x * 0 = x$, $x \wedge 0 = 0 \wedge x = 0$. In general, $x \wedge y \neq y \wedge x$.

Theorem 2.7 [10]

In any BCK- algebra, we have:

$$x * (y \wedge x) = x * y .$$

Proof:

Since $y \wedge x = x * (x * y) \leq y$

By Theorem 2.5 (1) $x * y \leq x * (y \wedge x)$

On the other hand, by BCK-2` we have:

$x * (y \wedge x) = x * (x * (x * y)) \leq x * y$

This means that: $x * (y \wedge x) = x * y$.

3. Subalgebras

Definition 3.1 [18]

A nonempty subset S of a BCK-algebra X is called a BCK-subalgebra of X, $if\ x * y \in S$ whenever $x, y \in S$.

Theorem 3.2 [10]

$C(x)$ is the intersection of subalgebras of X is a subalgebra of X.

Proof:

$$C(x) = \bigcap_i s_i \ ; s_i \ are\ subalgebras \text{ of } X$$

$$Let\ x, y \in C(x) = \bigcap_i s_i$$

$\Rightarrow x, y \in s_i \ \forall i$

$\because S\ subalgebra \Rightarrow x * y \in s_i \ \forall\ i$

$$\Rightarrow x * y \in \bigcap_i s_i = C(x)$$

$\Rightarrow C(x)$ is a subalgebra of X.

Remark 3.3

The union of two subalgebra of X is not necessarily to be a subalgebra of X.

Theorem 3.4 [10]

Let $(X\ ; *\ , 0)$ be a Bck-algebra and S is a subalgebra of X Then:

a) $0 \in S$.

b) $(S\ ; *\ ,0)$ is also a BCK – algebra.

c) X is a subalgebra of itself.

d) $\{o\}$ is a subalgebra of X.

Proof: Obvious.

Theorem 3.5 [10]

If we have a nonzero element x_0, of a BCK-algebra $(X\ ;\ *, 0)$, then $(\{0,\ x_0\};\ ,0)$ is a subalgebra of X.

Proof: Immediate.

Example 3.6 [7]

Let $Z = \{0, a, b, c\}$ and define a binary operation $*$ on Z by the following table:

$*$	0	a	b	c
0	0	0	0	0
a	a	0	a	a
b	b	b	0	b
c	c	c	c	0

It is clear that: $Z = (Z,*\ ,0)$ is a BCK-algebra. Then, $\{0\}, \{0, a\}, \{0, b\}, \{0, c\}, \{0, a, b\}, \{0, a, c\}, \{0, b, c\}$ and Z are all subalgebras of Z.

Definition 3.7 [18]

Let $(X,*\ ,0)$ be a BCK- algebra. For any $x\ , y\ in\ X$, we define

$$[x\ , y] = (x \wedge y) * (y \wedge x)$$

Example 3.8 [18]

Let $X = \{0, a, 1\}$ and $(*)$ operation be given by the following table:

$*$	0	a	1
0	0	0	0
a	a	0	0
1	1	a	0

Then $(X,*\ ,0)$ is a BCK-algebra and $A = \{0,1\}, B = \{0, a\}$ are two sub-algebras of X such that $[A, B] = [B, A] = \{0\}$.

Example 3.9 [18]

Let $X = \{0,1,2,3,4\}$ and let the $*$ operation be defined by the following table:

$*$	0	1	2	3	4
0	0	0	0	0	0
1	1	0	0	0	0
2	2	2	0	2	0
3	3	1	1	0	1
4	4	4	4	4	0

Then $(X, *, 0)$ is a BCK-algebra. $A = \{4\}$ and $B = \{2\}$ are two subsets of X. However, $[A, B] = \{[4,2]\} = \{2\}$ because $[4, 2] = (4 \wedge 2) * (2 \wedge 4) = (2 * (2 * 4)) * (4 * (4 * 2)) = (2 * 0) * (4 * 4) = 2 * 0 = 2$

Therefore, $2 \in [A, B]$ and $2 * 2 = 0$ is not member of $[A, B]$, i.e $[A, B]$ is not sub-algebra of X.

4. Bounded BCK-algebra

Definition 4.1 [11]

Let X be a BCK-algebra. Then, X is called bounded BCK-algebra if there exists the greatest element 1 of X, and for any $x \in X$, $1 * x$ is denoted by Nx.

Proposition 4.2 [11]

Let X be a bounded BCK-algebra. Then, for all $x, y \in X$

(i) $N1 = 0$ and $N0 = 1$

(ii) $N x * N y \leq y * x$

(iii) $N x * y = N y * x$

(iv) $NNx \leq x$ that $NNx = N(Nx)$,

(v) $y \leq x$ implies $Nx \leq Ny$

(vi) $NNNx = Nx$

(vii) $1 \wedge x = x$

(viii) $x \wedge 1 = NNx$

Example 4.3 [11]

Let $X = \{0, a, b, 1\}$, where $0 \leq a \leq b \leq 1$. The operation $*$ on X is defined as follows:

$*$	0	a	b	1
0	0	0	0	0
a	a	0	0	0
b	b	b	0	0
1	1	1	1	0

Then, $(X, *, 0)$ is bounded BCK-algebra with largest element 1.

5. Commutative BCK- algebras

Definition 5.1 [18]

A BCK-algebra X is said to be commutative if it satisfies:

$$x \wedge y = y \wedge x, \quad for\ all\ x, y \in X.$$

Properties 5.2 [18]

In any commutative BCK-algebra, the following statements hold:

$x \wedge x = x \vee x = x,$

$x \vee 0 = 0 \vee x = x \wedge 1 = 1 \wedge x = x,$

$x \wedge y = y \wedge x,$

$x \vee y = y \vee x,$

$x \vee 1 = 1 \vee x = 1,$

$0 \wedge x = x \wedge 0 = 0.$

Corollary 5.3 [18]

Let $(X, *, 0)$ be a commutative BCK-algebra. If A and B are two subsets of X, then $[A, B] = [B, A] = \{0\}$.

Example 5.4 [9]

Let $X = \{0,1,2,3\}$ be a BCK-algebra and the operation $*$ on X is defined as follows:

*	0	1	2	3
0	0	2	2	2
1	2	1	3	1
2	2	1	1	3
3	2	3	1	1

X is not commutative BCK-algebra because:

$$0 \wedge 2 \neq 2 \wedge 0$$

$$2 * (2 * 0) \neq 0 * (0 * 2)$$

$$2 * 2 \neq 0 * 2$$

$$1 \neq 2$$

Example 5.5 [18]

Let the $X = \{0 , a , 1\}$ and $(*)$ operation be given by the following table:

$*$	0	a	1
0	0	0	0
a	a	0	0
1	1	a	0

Then, $(X,* ,0)$ is a commutative bounded BCK-algebra with the largest element 1. Then $A = \{0,1\}$, $B = \{0, a\}$ are two subalgebras of X and $[A, B] = [B, A] = \{0\}$.

Theorem 5.6 [10]

For a BCK-algebra $(X,* ,0)$, the following are equivalent:

(a) X is a commutative,

(b) $x * (x * y) \leq y * (y * x)$,

(c) $\big(x * (x * y)\big) * \big(y * (y * x)\big) = 0$

6. Positive implicative BCK-algebra

Definition 6.1 [18]

BCK-algebra X is called implicative if and only if:

$$x * (y * x) = x.$$

Example 6.2[10]

Let $X = \{0,1\}$

$*$	0	1
0	0	0
1	1	0

X is an implicative BCK-algebra.

Example 6.3 [10]

Let $X = \{0, a, b, c, d, 1\}$

$*$	0	a	b	c	d	1
0	0	0	0	0	0	0
a	a	0	a	a	0	0
b	b	b	0	0	0	0
c	c	c	b	0	b	0
d	d	b	a	a	0	0
1	1	c	d	a	b	0

$d * (c * d) = d * b = a \neq d$. Then, $(X,*,0)$ is not implicative BCK-algebra

Definition 6.4 [18]

X is called a positive-implicative BCK-algebra if it satisfies in property:

$$(x * z) * (y * z) = (x * y) * z.$$

Example 6.5 [18]

Let $X = \{0,1,2,3,4\}$. Define $*$ by the following table:

*	0	1	2	3	4
0	0	0	0	0	0
1	1	0	1	0	0
2	2	2	0	0	0
3	3	3	3	0	0
4	4	3	4	1	0

Then, $(X,* ,0)$ is a bounded positive implicative BCK-algebra with the largest element 4.

Definition 6.6 [19]

A BCK-algebra $(X; *, 0)$ is called to be negative implicative if it satisfies:

$$(z * x) * (z * y) = z * (x * y) \quad \text{for all } x, y, z \text{ in } X$$

Corollary 6.7 [19]

A positive implicative BCK-algebra is not a negative implicative BCK–algebra.

Example 6.8 [19] Let $X = \{0, a, b, 1\}$ *and* $*$ *on* X be given by the table

*	0	a	b	1
0	0	0	0	0
a	a	0	a	0
b	b	b	0	0
1	1	1	1	0

Then, $(X; *, 0)$ *is* a positive implicative BCK–algebra. But, it is not negative implicative BCK-algebra as:

$$(a * 0) * (a * b) = a * a = 0$$

$$a * (0 * b) = a * 0 = a$$

$$(a * 0) * (a * b) \neq a * (0 * b)$$

Theorem 6.9 [10]

A BCK-algebra is implicative iff it is both commutative and positive implicative.

i.e., Implicative ↔ commutative + positive implicative

Proof:

"$\Longrightarrow$ " Let $(X, *, 0)$ be an implicative BCK-algebra. Then,

$x * y = (x * y) * [y * (x * y)] = (x * y) * y$

Hence, X is a positive implicative

By the positive implicative and BCK-1`,

$\therefore x * (x * y) = (x * (y * x)) * (x * y) \leq y * (y * x)$ ....(1)

by interchanging x and y in (1):

$y * (y * x) \leq x * (x * y)$ ....(2)

$\therefore x * (x * y) = \big(y * (y * x)\big)$

Then, X is commutative.

"$\Longleftarrow$ " Let X be both commutative and positive implicative

$x * [x * (y * x)]$

$= (y * x) * [(y * x) * x]$ by commutative

$= [(y * x) * x] * [(y * x) * x]$ by positive implicative

$= 0$ by BCI-3

$\therefore x * (y * x) = x$

X is implicative.

7. Ideals of a BCK-algebra

Definition 7.1 [10]

Let $(X,*,0)$ be a BCK-algebra and I be a nonempty subset of X. Then, I is said to be an ideal of X if, for all $x, y \in X$:

(i) $0 \in I$

(ii) $x * y \in I$ and $y \in I$ imply $x \in I$.

Example 7.2 [10]

$\{0\}$ and X trivial ideal of X.

Example 7.3 [10]

Let $X = \{0,1,2\}$

*	0	1	2
0	0	0	0
1	1	0	1
2	2	2	0

X is an implicative BCK-algebra and $\{0\}, X, \{0,1\}, \{0,2\}$ are ideals of X.

Example 7.4 [10]

Let $X = \{0,1,2,3\}$

*	0	1	2	3
0	0	0	0	0
1	1	0	0	0
2	2	1	0	0

3 3 2 1 0

X is commutative BCK-algebra it has only two ideals $\{0\}$ *and* X. {0,1}, {0,2}, {0,3}, {0,1,3}, {0,1,2}, {0,2,3} are not ideals of X.

Theorem 7.5 [10]

Suppose I is an ideal of Bck-algebra X and $x \in I$. If $y \leq x \rightarrow y \in I$

Proof:

$$y \leq x \rightarrow y * x = 0 \in I \text{ and } x \in I \rightarrow y \in I$$

Theorem 7.6 [10]

Suppose $I \subsetneq X$. Then I is an ideal iff $\forall x, y \in I$,

$$z * x \leq y \rightarrow z \in I$$

$i.e.\ (z * x) * y = 0 \rightarrow z \in I$

Proof:

" $\Rightarrow$ " let I ideal, $(z * x) * y = 0 \in I$ and $y \in I \rightarrow z * x \in I$ and

$$x \in I \rightarrow z \in I$$

" $\Leftarrow$ " $0 = (0 * x) * y \rightarrow 0 \in I$

Let $x * y \in I$, $y \in I$

$$(x(x * y)) * y = 0, y \in I \rightarrow x * (x * y) \in I$$

and $x * y \in I \rightarrow x * I$ and I ideal

Corollary 7.7 [10]

I is an ideal $\Leftrightarrow \forall x, y \in I, (z * x) * y = 0 \rightarrow z \in I$

Theorem 7.8 [10]

Any ideal of X is a subalgebra of X.

Proof :

Let I is an ideal of X and $x, y \in I$. Since $x * y \leq x$ By Theorem 7.5, $x * y \in I$. Hence, I is a subalgebra of X.

Theorem 7.9 [20]

Let $(X,*,0)$ be a BCK-algebra and I be an ideal of it. Then, $({}^{X}/_{I},*,I)$ is also a BCK-algebra, which is called to be the quotient algebra via I.

Definition 7.10 [20]

A nonempty subset I of X is called:

- A commutative ideal of X, if $0 \in I$ and if $(x * y) * z \in I$ and $z \in I$ imply that $x * (y * (y * x)) \in I$, for all $x, y, z \in X$.

- An implicative ideal of X, if $0 \in I$ and if $(x * (y * z)) * z \in I$ and $z \in I$ imply that $x \in I$, for all $x, y, z \in X$.

- A positive implicative ideal of X, if $0 \in I$ and if $(x * y) * z \in I$ and $y * z \in I$ imply that $x * z \in I$, for all $x, y, z \in X$.

- A maximal ideal of X, if I is a proper ideal of X and not a proper subset of any proper ideal of X.

- An obstinate ideal of X, if I is a proper ideal of X and if $x, y \notin I$ implies $x * y \in I$ and $y * x \in I$, for all $x, y \in X$.

- An irreducible ideal of X, if I is an ideal of X and if $I = J \cap K$ implies $I = J$ or $I = K$, for ideals J, K of X.

- A normal ideal of X, if I is an ideal of X and $x * (x * y) \in I$ implies $y * (y * x) \in I$, for all $x, y \in X$.

Definition 7.11 [21]

In what follows let X and $\mathbb{N}$ denote a BCK-algebra and the set of all positive integers, respectively, unless otherwise species. For any x and y of a BCK-algebra X, let $x * y^k$ denote

$(\ldots((x * y) * y) * \ldots) * y$ in which y occurs k times, where $k \in \mathbb{N}$.

Definition 7.12 [21]

For any $a, b \in X$ and $k \in \mathbb{N}$ we define

$$[a; b^k] := \{x \in X \mid (x * a) * b^k = 0\}$$

Obviously , $0, a, b \in [a; b^k]$ $for\ all\ a, b \in X\ and\ k \in \mathbb{N}$.

Theorem 7.13 [21]

If X is positive implicative, then $[a;\ b^k]$ is an ideal of X $for\ all\ a, b \in X$ and $\in \mathbb{N}$.

Proof :

Let $x, y \in X$ be such that $x * y \in [a;\ b^k]$ and $y \in [a;\ b^k]$. Then

$$0 = ((x * y) * a) * b^k = (((x * a) * (y * a)) * b) * b^{k-1}$$

$$= \Big(\big((x * a) * b\big) * \big((y * a) * b\big)\Big) * b^{k-1}$$

....

$$= ((x * a) * b^k) * ((y * a) * b^k)$$

$$= ((x * a) * b^k) * 0 = (x * a) * b^k$$

And so $\in [a; b^k]$. Therefore, $[a;\ b^k]$ is an ideal of X.

Theorem 7.14 [21]

Let I be a non-empty subset of X. Then I is an ideal of X if and only if $[a; b^k] \subseteq I$ for every $, b \in I \ and \ k \in \mathbb{N}$.

Proof:

Assume that I is an ideal of X and let a, b $\in I \ and \ k \in \mathbb{N}. if \ x \in [a;\ b^k]$,

Then $(x * a) * b^k = 0 \in I$. Since, $a, b \in I$, by using (definition 7.1 (ii)) repeatedly we get $x \in I$. Hence $[a; b^k] \subseteq I$.

Conversely, suppose that $[a; b^k] \subseteq I$ for all $a, b \in I \ and \ k \in N$. Note that $0 \in [a; b^k] \subseteq I$.

Let $x, y \in X$ be such that $x * y \in I \ and \ y \in I$. Then

$$(x * (x * y)) * y^k = ((x * (x * y)) * y) * y^{k-1} = 0 * y^{k-1} = 0$$

and thus $x \in [x * y; y^k] \subseteq I$. Hence I is an ideal of X.

Example 7.15 [21]:

Consider a BCK-algebra $X = \{0, a, b, c, d\}$ with the following Cayley table.

*	0	a	b	c	d
0	0	0	0	0	0
a	a	0	a	0	0
b	b	b	0	b	0
c	c	a	c	0	a
d	d	d	d	d	0

Then $[d;\ b^2] = \{0, a, b\ , d\ \}$ is not an ideal of X because $c\ *\ d\ =\ a\ \in [d; b^k]$ and $d\ \in [d; b^k]$, but $c\ \notin [d;\ b^2]$.

-We use the notation $x\ \wedge\ y$ instead of $y *\ (y\ * x)$ for all $, y\ \in X$.

Definition 7.16 [21]

A non-empty subset I of X is called a quasi-left (resp. quasi-right) ideal of X if

(1)$0\ \in\ I$,
(2) $x\ \in\ I$ and $y\ \in\ X$ imply $y\ \wedge\ x\ \in\ I$ (resp. $x\ \wedge\ y\ \in\ I$).

If an ideal I is both quasi-left and quasi-right, we say that I is a quasi-ideal.

Theorem 7.17 [21]

For every $a\ , b\ \in\ X$ and $k\ \in\ N$, the set $[a;\ b^k]$ is a quasi-ideal of X.

Proof: Note that $0 \in [a; b^k]. Let\ x\ \in [a, b^k] and\ y \in X$. Then

$((y \wedge x) * a) * b^k = ((x * (x * y)) * a) * b^k$

$= \big((x * a) * (x * y)\big) * b^k$

$= \Big((x * a) * b^k\Big) * (x * y)$

$=\ 0 * (x * y) = 0$

And so $y\ \wedge x \in [a; b^k]$. Therefore $[a; b^k]$ is a quasi- left ideal of . Now we get

$$\big((x\ \wedge y) * a\big) * b^k = \Big(\big(y * (y * x)\big) * a\Big) * b^k$$

$$= \big((y * a) * (y * x)\big) * b^k$$

$$\leq (x * a) * b^k = 0$$

And hence $((x\ \wedge y\ ) * a) * b^k\ =\ 0$ which shows that $\wedge\ y \in [a; b^k]$, i.e., $[a; b^k]$ is quasi- right ideal of X . This completes the proof.

Theorem 7.18 [21]

If I is an ideal of X, then $I = \bigcup_{a,b\in I}[a; b^k]$ for every $k \in N$.

Proof:

Let I be an ideal of X. Then the inclusion $\bigcup_{a,b\in I}[a; b^k] \subseteq I$ follows from Theorem 7.14

Let $x \in I$. Since, $x \in [x;\ 0^k]$, it follows that

$$I \subseteq \bigcup_{x\in I}[x; 0^k] \subseteq \bigcup_{a,b\in I}[a; b^k]$$

This completes the proof.

Definition 7.19 [10]

A node of X is an element of X which is comparable with every element of X. It is clear that 0 is a node in every BCK-algebra.

Proposition 7.20 [20]

An element $a \in X$ is a node if and only if for every $x \in X$ either $a * x = 0$ or $x * a = 0$.

Example 7.21 [20]

a) Let $X = \{0, a, b, c\}$. For all $x, y \in X$, we define, $*$ as follows:

*	0	a	b	c
0	0	0	0	0
a	a	0	0	a
b	b	a	0	b
c	c	c	c	0

Then, (X,$*$, 0) is a BCK-algebra. 0 is only node of X

b) Let $X = \{0,1,2,3,4\}$. For all $x, y \in X$, we define, $*$ as follows:

*	0	1	2	3	4
0	0	0	0	0	0
1	1	0	0	0	0
2	2	1	0	1	0
3	3	3	3	0	0
4	4	4	4	4	0

Then, $(X,*,0)$ is a BCK-algebra. $\{0,1,4\}$ is the set of all nodes of X.

c) Let $\mathrm{X} = \{0, \mathrm{a}, \mathrm{b}, 1\}$. For all $x, y \in X$, we define, $*$ as follows:

*	0	a	b	1
0	0	0	0	0
a	a	0	0	0
b	b	b	0	0
1	1	b	a	0

Then $(X,*,0)$ is a BCK-algebra. All elements of X are node

d) Let $\mathrm{X} = \{0, \mathrm{a}, \mathrm{b}, 1\}$. For all $x, y \in X$, we define, $*$ as follows:

*	0	a	b	1
0	0	0	0	0
a	a	0	a	0
b	b	b	0	0
1	1	1	1	0

Then, $(X,*,0)$ is a BCK-algebra and $\{0,1\}$ is the set of all nodes of X.

Definition 7.22 [20]

An ideal I of X will be called a nodal ideal of X, if I is a node of $I(X)$. We denote all nodal ideals of X by $N(X)$.

Example 7.23 [20]

a) X and $\{0\}$ are trivial nodal ideals of every X.

b) In Example 7.21(a), we have $I(X) = \{\{0\}, \{0, c\}, \{0, a, b\}, X\}$. But only $\{0\}$ and X are nodal ideals of X.

c) In Example 7.21(b), $\{0\}, \{0, 1, 2\}, \{0, 1, 2, 3\}, X$ are all of nodal ideals of X.

d) In Example 7.21(c), $\{0\}, \{0, a\}, X$ are all of nodal ideals of X.

e) In Example 7.21(d), we have $I(X) = \{\{0\}, \{0, a\}, \{0, b\}, \{0, a, b\}, X\}$. And $\{0\}, \{0, a, b\}, X$ are all of nodal ideals of X.

Theorem 7.24 [20]

Let I be an ideal of X. If for all $x, y \in X$ such that $x \in I$ and $y \notin I$, the relation $x < y$ is satisfied, then I is a nodal ideal of X.

Proof: If I is not a nodal ideal of X. So there exists an ideal J of X such that J incomparable with I. Then there are two elements $x, y \in X$ such that $x \in I - J$, $y \in J - I$ and $x \nless y$. Thus it is contrary, so every ideal J of X is comparable with I, that is, I is a nodal ideal of X

Theorem 7.25 [20]

If I and J are two nodal ideals of X, then

(i) $I \cap J$ *is a nodal ideal of* X,
(ii) $I \cup J$ *is a nodal ideal of* X.

Definition 7.28 [5]
A nonempty subset $I^{\vee}$ of bounded BCK-algebra X is said to be a dual ideal of X if
$(i) 1 \in I^{\vee}$
$(ii) N(Nx * Ny) \in I^{\vee}$ and $y\ I imply\ x \in I^{\vee}, for\ any\ x, y\ X$
Example 7. 29 [5]

$Let\ X = \{0, a, b, c, 1\}$ and the operation $*$ is given by the following table :

$*$	0	a	b	c	1
0	0	0	0	0	0
a	a	0	0	0	0
b	b	a	0	a	0
c	c	c	c	0	0
1	1	c	c	a	0

Therefore, $(X,*,0)$ is a bounded BCK-algebra, $I^{\vee} = \{c, 1\}$ is a dual ideal of X.

7.1 Positive Implicative Ideals

Definition 7.1.1 [10]

An ideal I of X is called a positive implicative ideal if

$(i) 0 \in I$

$(ii) x, y, z \in X, (x * y) * z \in I, y * z \in I \rightarrow x * z \in I$

Example 7.1.2 [10]

Let $X = \{0,1,2,3,4\}$ in which the operation $*$ is given by the table:

*	0	1	2	3	4
0	0	0	0	0	0
1	1	0	1	0	1
2	2	2	0	2	0
3	3	1	3	0	3
4	4	4	4	4	0

It is easy to verify that $\{0,1,3\}, \{0,1,2,3\}$ are positive implicative ideals of X. Also, $\{0\}, \{0,2\}, \{0,2,4\}$ are ideals but not positive implicative.

Theorem 7.1.3 [10]

Any positive implicative ideal is an ideal of X but the inverse is not true in general.

Proof: suppose I positive implicative ideal of X.

Let $x * y \in I , y \in I \rightarrow (x * y) * 0 \in I$

$\because y \in I \rightarrow y * 0 \in I$

Since I positive implicative ideal, then $x * 0 = x \in I$

Hence, I is an ideal of .

Theorem 7.1.4 [10]

I is a positive implicative ideal $\Leftrightarrow A_a = \{x \in X; x * a \in I\}$ is an ideal of X for any $a \in X$

Proof:

$\Rightarrow$" Let I be a positive implicative ideal and $x * y \in A_a$, $y \in A_a$ "

Then, $(x * y) * a \in I$ and $y * a \in I$

Since I positive implicative ideal, then $x * a \in I \rightarrow x \in A_a$

Hence, A_a is an ideal of X.

" $\Leftarrow$" Let A_a be an ideal of X and $(x * y) * z \in I, y * z \in I$

$\rightarrow x * y \in A_z, y \in A_z$

$\because A_z$ is an ideal of X, then $x \in A_z \rightarrow x * z \in I$

$\therefore I$ is a positive implicative ideal of X.

Corollary 7.1.5 [10]

If I is a positive implicative ideal of X, then A_a is the least ideal of X containing I and a.

Proof:

Let B be any ideal containing I and a

Let $x \in A_a \rightarrow x * a \in I, I \subseteq B$

$\therefore x * a \in B, a \in B \rightarrow x \in B$

$\therefore A_a \subseteq B$

$\therefore A_a$ the least ideal containing I and a.

Theorem 7.1.6 [10]

Given a nonempty subset I of a BCK-algebra X, then I is positive implicative ideal iff $(x * y) * z \in I \Longrightarrow (x * z) * (y * z) \in I \quad \forall x, y, z \in I$.

Corollary 7.1.7 [10]

In a BCK- algebra X, the following are equivalent:

(a) X is a positive implicative BCK-algebra,
(b) $\{0\}$ is a positive implicative ideal of X,
(c) All ideals of X is a positive implicative,
(d) For any a in X, $A_a = \{x \in X; x * a \in I\}$ is an ideal of X.

7.2 Maximal ideals and prime ideals

Definition 7.2.1[8]

Given a BCK-algebra $(X,* ,0)$, an ideal I of X is called a maximal ideal if I is a proper ideal of X and not a proper subset of any proper ideal of X. An ideal I is proper if $I \neq X$.

Example 7.2.2 [8]

Let $X = \{0,1,2,3,4\}$, in which $*$ is given by the table:

$*$	0	1	2	3	4
0	0	0	0	0	0
1	1	0	0	0	0
2	2	1	0	1	1
3	3	3	3	0	3
4	4	4	4	4	0

Then $(X,* ,0)$ is a BCK-algebra. It is easy to verify that $\{0,1,2,3\}$ and $\{0,1,2,4\}$ are two maximal ideal of X.

Definition 7.2.3 [6]

An ideal I in a commutative BCK-algebra X is called prime if:

$x \wedge y \in I \textit{ implies } x \in I \textit{ or } y \in I.$

Theorem 7.2.4 [6]

If M is a maximal ideal of BCK-algebra X, then M is a prime ideal of X.

Example 7.2.5 [10]

Let $= \{0, 1, 2 ,3 ,4 \}$, in which $*$ is given by the table:

*	0	1	2	3	4
0	0	0	0	0	0
1	1	0	0	0	0
2	2	1	0	1	0
3	3	3	3	0	0
4	4	4	4	4	0

$I = \{0,1,2\}$ prime ideal of X but not maximal ideal as there exist $J = \{0,1,2,3\}$ is an ideal such that $I \subseteq J$.

Definition 7.2.6 [3]

Let A be a BCK-algebra. A nonempty subset S of A is called A-closed if and only if $x \wedge y \in S$ whenever $x, y \in S$.

Proposition 7.2.7 [3]

Let A be a bounded commutative BCK-algebra. An ideal P of A is prime If and only if $S = A\backslash P$ is a A-closed subset of A. Moreover S is a ideal of A if P is prime ideal.

(1)Assume P is a prime ideal of A. Let $x, y \in S$. If $x \wedge y \notin S$, then $x \wedge y \in P$. Since P is a prime ideal, then either $x \in P$ or $y \in P$. This contradicts the assumption that x and y both belong to S. Hence $x \wedge y \in S$. So, S is A-closed.

(2) Conversely, assume that $S = A\backslash P$ is a A-closed subset of A. Suppose $x \wedge y \in P$. If $x \notin P$ and $y \notin P$, then $x, y \in S$. So, $x \wedge y \in S$. since S is A-

closed. This is a contradiction hence either $x \in P$ or $y \in P$, showing that P is a prime ideal. Finally, it is clear that S is a ideal of A if P is a prime ideal of A.

7.3 Implicative ideals

Definition 7.3.1 [10]

Let $(X,*,0)$ be a BCK-algebra. A non- empty subset I of X is said to be an implicative ideal if it satisfy the following:

(i) $0 \in I$
(ii) If $(x*(y*x))*z \in I, z \in I \Rightarrow x \in I$

Example 7.3.2 [10]

X is always an implicative ideal of itself, which is called the trivial implicative ideal.

Example 7.3.3 [10]

Let X be an implicative BCK-algebra. Then every ideal of X is an implicative ideal

Solution: Let I be an ideal of and $(x*(y*x))*z \in I, z \in I$.

Since I is an ideal of X, then $0 \in I$ from Definition 7.1 (i)

Since $(x*(y*x))*z \in I, z \in I, I$ is an ideal, then $x*(y*x) \in I$

Since X is an implicative BCK-algebra, then $x = x*(y*x) \in I$.

$\therefore$ I is an implicative ideal of X.

Example 7.3.4 [10]

Let $X = \{0,,2,3,4\}$ in which the operation $*$ is given by the table:

$*$	0	1	2	3	4
0	0	0	0	0	0
1	1	0	0	0	0
2	2	1	0	1	0
3	3	3	3	0	0
4	4	4	4	4	0

It is easy to verify that $I = \{0,1,2,3,4\}$ is an implicative ideal of X.

Theorem 7.3.5 [10]

An implicative ideal must be an ideal, but the inverse is not true in general.

Proof. Let I be an implicative ideal and $x * z \in I, z \in I$

Let $y = x$ in (ii) $(x * (x * x)) * z = x * z \in I, z \in I$

Since I is an implicative ideal, then $x \in I$ and hence I is an ideal but the inverse is not true by the following example.

Example 7.3.6 [10]

Let $= \{0,1,2,3,4\}$, the operation $*$ is given by the table

$*$	0	1	2	3	4
0	0	0	0	0	0
1	1	0	1	0	1
2	2	2	0	2	0
3	3	1	3	0	3
4	4	4	4	4	0

Then $(X,*,0)$ is a BCK-algebra. It is easy to verify that $I = \{0,2\}$ is an ideal but not implicative.

Theorem 7.3.7 [10]

An implicative ideal must be positive implicative ideal, but the converse is not true in general.

Proof: suppose that I be an implicative ideal of X.

Let $(x * y) * z \in I, y * z \in I$

$\therefore \big((x * z) * z\big) * (y * z)) \leq (x * z) * y = (x * y) * z \in I$

$\rightarrow \big((x * z) * z\big) * (y * z) \in I, y * z \in I$

$\rightarrow (x * z) * z \in I$

Since $\Big((x * z) * \big(x * (x * z)\big)\Big) * 0 \in I, 0 \in I$

$\because I$ is an implicative ideal, then $x * z \in I$

$\therefore I$ positive implicative ideal of X

But, the converse is not true as shown by the following example:

Example 7.3.8 [10]

Let $X = \{0,1,2,3,4\}$. The operation $*$ is given by the table:

$*$	0	1	2	3	4
0	0	0	0	0	0
1	1	0	1	0	1
2	2	2	0	2	0
3	3	1	3	0	3
4	4	4	4	4	0

It is easy to verify that $\{0,1,3\}$ is positive implicative ideal but not implicative ideal.

7.4 Normal ideal

Definition 7.4.1 [4]

An ideal $A\ of\ X$ is called a normal ideal if $x * (x * y) \in A$ implies $y * (y * x) \in A\ for\ all\ x, y \in X.$

Example 7.4.2 [4]

Let $(X = \{0,1,2,3\}, * ,0)$ be a BCK-algebra with the following Cayley table:

*	0	1	2	3
0	0	0	0	0
1	1	0	1	0
2	2	2	0	0
3	3	2	1	0

It is easy to verify that $C = \{0,1\}$ is a normal ideal of X.

Definition 7.4.3 [4]

For a bounded BCK-algebra X, if an element $x\ in\ X$ satisfies $NNx = x$, then x is called an involution. If any element in X is an involution, then X is called an involutory BCK-algebra.

Definition 7.4.4 [4]

$If\ X$ is a BCK-algebra and A a non-empty subset of X, then the set $A^* := \{x \in X | (\forall a \in A)\ a * (a * x) = 0\}$ is called the annihilator of A.

Definition 7.4.5 [4]

Let X be a bounded BCK-algebra and A, C non-empty subsets of X. Then, the set $(C : A)^d = \{x \in X \mid Na * (Na * Nx) \in C \; for \; all \; a \in A\}$ is called the dual annihilator of A with respect to C.

Definition 7.4.6 [1]

$A^{**} = (A^*)^*$ is the double annihilator of $A . if \; A = A^{**}$ then A is called an involutory ideal.

Definition 7.4.7 [1]

For an ideal A, $x^{-1} A = X$ if and only if $x \in A$.

Lemma 7.4.8 [4]

Let X be a bounded BCK-algebra, C an ideal of $X \; and \; A \subseteq X. Then \; NC \subseteq (C : A)^d, where \; NC = \{1 * c \mid c \in C\}$.

Definition 7.4.9 [4]

A non-empty subset D of a bounded BCK-algebra X is said to be a dual ideal of X if:

(1)$1 \in D$;

(2) $N(Nx * Ny) \in D \; and \; y \in D \; imply \; x \in D \; for \; any \; x, y \in X$.

Proposition 7.4.10 [4]

Let X be an involutory BCK-algebra, C an ideal and $A \subseteq X$. Then the following are hold:

(i) if $(C : A)^d = X, then\ A \subseteq NC$;
(ii) if in addition C is normal, $then\ (C : A)^d = X\ if\ and\ only\ if\ A \subseteq NC.$

Proof:

(i) Let $(C : A)^d = X$ and $x \in A$.

Then $x \in (C : A)^d$ and so $Na * (Na * Nx) \in C\ for\ every\ a \in A.$

Thus from $x \in A, we\ get\ Nx * (Nx * Nx) \in C$, that is, $Nx \in C$. It follows that $NNx \in NC$. But by the involutory property of X, we have $NNx = x$. Therefore $x \in NC\ and\ so\ A \subseteq NC.$

(iii) By (i), we only need to prove the sufficiency. Let $A \subseteq NC$ and let x be an arbitrary element of X. Using axiom BCI-2, we have:
$Nx * (Nx * Na) \leq Na \in NA\ for\ any\ a \in A.$ (**)

Now, for any $a \in A$, by hypotheses, there exists $c \in C\ such\ that\ a = Nc.$ Thus by (**), we get $Nx * (Nx * Na) \leq NNc. But\ NNc \leq c\ for\ any\ c \in C$. Hence $Nx * (Nx * Na) \in C$ and so by the normality of C, we conclude $Na * (Na * Nx) \in C$, which implies $x \in (C : A)^d$. Therefore $(C : A)^d = X$, and so the proof is completed.

Theorem 7.4.11 [4]

Let X be an involutory BCK-algebra. Then for any normal ideal C and two dual ideals $A, B\ of\ X,$

$A \cap B \subseteq NC \Leftrightarrow A \subseteq (C{:}B)^d$

Theorem 7.4.12 [4]

Let X be a bounded BCK-algebra, C a normal ideal of X and $A \subseteq X$. Then the following are hold:

(i) $A \subseteq (C : (C : A)^d)^d$;
(ii) $(C : A)^d = (C : (C : (C : A)^d)^d)^d$.

7.5 Irreducible ideal

Definition 7.5.1 [2]

A proper ideal I of a BCK-algebra X is called to be irreducible if:

$I = A \cap B \; implies \; I = A \; or \; I = B \; for \; any \; A, B \; ideal \; of \; X.$

Theorem 7.5.2 [16]

In a BCK-algebra, the following conditions are equivalent

(i) P is an irreducible ideal؛

(ii) P is a prime ideal;

(iii) for any ideals $A, B, A \cap B \subseteq P \; implies \; A \subseteq P \; or \; B \subseteq P.$

7.6 Commutative ideal

Definition 7.6.1 [10]

A nonempty subset I of a BCK-algebra X is said to be a Commutative ideal of X if it satisfies :

(i) $0 \in I$

(ii) $(x * y) * z \in I, z \in I \rightarrow x * (y * (y * x)) = x * (x \wedge y) \in I, \forall x, y, z \in X$

Example 7.6.2 [10] X , $\{0\}$ are Commutative ideal

Example 7.6.3 [10]

Let $X = \{0,1,2,3,4\}$. The operation $*$ is given by the table:

$*$	0	1	2	3	4
0	0	0	0	0	0
1	1	0	1	0	1
2	2	2	0	2	0
3	3	1	3	0	3
4	4	4	4	4	0

It is easy to verify that $\{0,2\}$, $\{0,2,4\}$ are Commutative ideals.

Theorem 7.6.4 [10]

A Commutative ideal must be an ideal. But, the converse is not true in general.

Proof: suppose I be a Commutative ideal, $x * y \in I , y \in I$

Then $(x * 0) * y = x * y \in I , y \in I$

Since I is a Commutative ideal

$\rightarrow x * \big(0 * (0 * x)\big) = x * (0 * 0) = x * 0 = x \in I$

$\therefore I$ is an ideal of X.

Example 7.6.5 [10]

In Example 7.6.3, it is easy to verify that $\{0,1,3\}$ is an ideal but not commutative.

Theorem 7.6.6 [10]

An ideal I is commutative if and only if it satisfies:

$$x * y \in I \Rightarrow x * (y * (y * x)) \in I$$

8. Quotient of a BCK-algebra

Definition 8.1 [10]

Let $(X,*,0)$ be a Bck-algebra, x and $y \in X$, we define an equivalence relation $x \sim y$ if and only if $x * y \in I$ and $y * x \in I$ for some ideal I of X. The equivalent class C_x is defined as:

$C_x = \{y \in X\,; x{\sim}y\} = \{y \in X\,; x * y \in I\,, y * x \in I\}$

And so,

$C_0 = \{y \in X\,; 0{\sim}y\} = \{y \in X\,; 0 * y = 0 \in I\,, y * 0 = y \in I\} = I$

Definition 8.2 [10]

Let $X/_I = \{C_x \;|x \in X\,\}$ be the set of all equivalence classes. Define $*$ on $X/_I$ by:

$C_x * C_y = C_{x*y}\ \forall x, y \in X$

So, $C_0 * C_x = C_{0*x} = C_0$

$\therefore C_0 = I\ is\ the\ least\ element\ of\ X/_I$.

Theorem 8.3 [10]

$(X/_I\ ,*, C_0)\ is\ BCk - algebra.$

Proof:

Since $C_0 * C_x = C_{0*x} = C_0$

$\Rightarrow C_0 \leq C_x\ \forall\, x \in X \Rightarrow\ Bck - 5'$ hold.

Let $C_x * C_y = C_y * C_x = C_0$. Then, $x \leq y\,, y \leq x\ \Rightarrow x = y$ and

$x * y{\sim}0\ and\ y * x{\sim}0.$ If $u \in C_x \rightarrow x{\sim}u \rightarrow u * y \sim x * u \rightarrow u * y{\sim}0\ \rightarrow u{\sim}y$

$\rightarrow u \in C_y\ \rightarrow C_x \subseteq C_y \ldots \ldots (1)$

Similarly, $C_y \subseteq C_x \ldots \ldots (2) \Longrightarrow C_y = C_x\ \Rightarrow Bck - 4'$ hold.

$C_x * C_x = C_{x*x} = C_0 \Longrightarrow C_x \leq C_x\ \forall\, x \in X \Rightarrow Bck - 3'$ hold.

If $x \leq y \Rightarrow x * y = 0 \Rightarrow C_x * C_y = C_{x*y} = C_0 \Rightarrow C_x \leq C_y$

$$\Rightarrow (C_x * C_y) * (C_x * C_z) = C_{x*y} * C_{x*z} = C_{(x*y)*(x*z)}$$

$\leq C_{z*y} = C_z * C_y \Rightarrow BcI - 1'$ hold.

Since $C_x * (C_z * C_y) = C_x * C_{z*y} = C_{x*(z*y)} \leq C_y \Rightarrow BcI - 2'$ hold.

$\Rightarrow (X/_I ,*, C_0)$ is BCk – algebra is said to be quotient algebra.

Definition 8.4 [1]:

A commutative BCK-algebra X is cancellative if $x \wedge y = 0$ implies $x = 0\ or\ y = 0\ for\ x,\ y \in X\ , that\ is\ (x)^* = 0\ for\ all\ x \in X\ with\ x \neq 0.$

Definition 8.5 [1]:

Let X be a commutative BCK-algebra and let A be an ideal of X. Suppose that B is a subset of X. Then we define the set $(A : B) = \{x \in X : x \wedge B \subseteq A\}$ as the generalized annihilator of B (relative to A). We observe that if $A = \{0\}$, then $B^* = (0 : B)$ and $(A : B)$ is non-empty because 0e $(A: B)$.

Remark 8.6 [1]

One can observe that if $x \in (A : B)$, then $x \wedge B \subseteq A$ and hence $B \subseteq x^{-1}A$. This implies that $(A : B) = \{x \in X : B \subseteq x^{-1}A\}$.

Remark 8.7 [1]

Let A be an ideal of a BCK-algebra X. Consider the quotient BCK-algebra $X/_A$. If $J/_A$ is a subset of $X/_A$, then we have :

$({}^{J}/_{A})^* = \{C_x : C_x \wedge {}^{J}/_{A} = A\} = \{C_x : J \subseteq x^{-1} A\}.$

Now we discuss the annihilator of an element of ${}^{X}/_{A}$. Let $C_x \in {}^{X}/_{A}$. Then

$$(C_x)^* = \{C_y : C_x \wedge C_y = A\} = \{C_y : C_{x\wedge y} = A\} = \{C_y : x \wedge y \in A\} = \{C_y : y \in x^{-1}A\}.$$

If $x \in A$, then $x^{-1}A = X$ (Definition 7.4.7) and hence $(C_x)^* = {}^{X}/_{A}$. If A is a prime ideal of X and C_x is a non-zero element of ${}^{J}/_{A}$, then $x \notin A$ and hence $x^{-1} A = A$ (Definition 7.4.7). This implies that $(C_x)^* = A$ (the zero element of ${}^{X}/_{A}$). All these observations lead to the following result:

Proposition 8.8 [1]

Let A be an ideal of a BCK-algebra X, let ${}^{J}/_{A}$ be a subset of ${}^{X}/_{A}$ and C_x an element of ${}^{X}/_{A}$. Then, the following statements hold:

(i) $({}^{J}/_{A})^* = \{C_x : x \in (A : J)\} = \{C_x : J \subseteq x^{-1}A\}$,

(ii) $(C_x)^* = \{C_y : y \in x^{-1}A\}$,

(iii) If A is a prime ideal of X and $C_x \neq A$ (non-zero element of ${}^{X}/_{A}$), then $(C_x)^* = A\ (zero\ of\ {}^{X}/_{A})$,

(iv) $({}^{J}/_{A})^{**} = \{C_x : x \in (A : (A : J))\}$.

Corollary 8.9 [1]

If X is an involutory BCK-algebra, then every quotient BCK-algebra of X is an involutory BCK-algebra.

Definition 8.10 [10]

A BCK-algebra $(X; * ,0)$ is called to be simple if it has no proper non-zero ideals.

Example 8.11 [10]

Let $X = \{0,1,2,3,4\}$ in which $*$ is given by the table:

$*$	0	1	2	3	4
0	0	0	0	0	0
1	1	0	0	0	0
2	2	1	0	0	0
3	3	1	1	0	0
4	4	1	1	1	0

Then $(X;*,0)$ is a simple BCK-algebra.

Proposition 8.12 [1]

Let X be an involutory BCK-algebra. Then X is cancellative if and only if X is simple.

Corollary 8.13 [1]

If A is a prime ideal of a BCK-algebra X, then X/A is cancellative.

Observe that if X is a cancellative involutory BCK-algebra, then it is simple.

Indeed, if A is an ideal of X, then

$$A = A^{**} = \bigcap_{x \in A} * \, (x)^*$$

Since X is cancellative, therefore $(x)^* = \{0\}$ for all non-zero elements $x \in X$ and hence $A = \{0\}$ or $A = X$. This proves that X is simple.

Corollary 8.14 [1]

Let A be an ideal of involutory BCK-algebra X. Then $X/_A$ is simple if and only if A is prime.

Proof. Let A be a prime ideal of X. Then $X/_A$is a cancellative (Corollary 8.13) and involutory BCK-algebra (Corollary 8.9). This implies that $X/_A$ is simple (Proposition 8.12). Conversely, assume that $X/_A$ is simple. This implies that $X/_A$ is cancellative (Proposition 8.12). Let $x \wedge y \in A$. Then, $C_{x \wedge y} = A$, $C_x \wedge C_y = A$. Since $X/_A$ is cancellative, therefore $C_x = A \; or \; C_y = A$. Consequently, $x \in A \; or \; y \in A$ and this implies that A is prime. This completes the proof.

Theorem 8.15 [10]

If I and J are any ideals of X and $I \subset J$, then

(a) I is also an ideal of the subalgebra J.

(b) $J/_I$ as the quotient of the subalgebra J via the ideal is an ideal of $X/_I$.

Proof:

(a) is immediate from Definition 7.1 . In order to prove (b), first We have that each element of $J/_I$ is also an element of $X/_I$. To avoid the ambiguity, we denote the element of $J/_I$ containing x by $C_x(J)$. Suppose $y \in X$ and $x \in J$.

If $x \sim y$ with respect to I, then $y * x \in I$, and so $y * x \in J$ and $x \in J$. By Definition 7.1(ii) we have $y \in J$, this says $C_x(J) \in X/_I$ or each element of $J/_I$ is also an element of $X/_I$.

Next, we prove that $J/_I$ is an ideal of $X/_I$. Since$I \subseteq J$, $C_0 = I \in J/_I$.

If $C_x * C_y \in J/_I$ and $C_y \in J/_I$ then $C_{x*y} \in J/_I$. It follows that $x * y \in J$ and $\in J$. By Definition 7.1(ii), we obtain $x \in J$, and so $C_x \in J/_I$. This means that $J/_I$ is an ideal of $X/_I$. This finishes the proof.

Theorem 8.16 [10]

Let I be an ideal of a BCK-algebra X. Then I is commutative if and only if the quotient is a commutative BCK-algebra .

Proof:

Necessity: Let I be a commutative ideal.

For x, y in X denote $u = y * (y * x)$, then $u * x = 0 \in I,$ and so $u * (x * (x * u)) \in I$ Since $C_0 = I$ it follows that

$$C_u * \left(C_x * (C_x * C_u)\right) = C_{u*(x*(x*u))} = C_0,$$

Hence,

$$C_y * \left(C_y * C_x\right) = C_{y*(y*x)} = C_u \leq C_x * (C_x * C_u)$$

The inverse inequality is trivial. Therefore, we have

$$C_y * \left(C_y * C_x\right) = C_x * (C_x * C_u) = C_x * \left(C_x * (C_y * \left(C_y * C_x\right))\right).$$

By Theorem 5.2 it follows that the quotient algebra ($X/_I ; *, C_0$) is commutative.

Sufficiency. Let ($X/_I : *, C_0$) be commutative, then zero ideal $\{C_0\}$ is commutative. If $x * y \in I$ then,

By Theorem 7.6.6, we have.

$$C_{x*(y*(y*x))} = C_x * \left(C_y * (C_y * C_x)\right) \in \{C_0\}$$

That is, $C_{x*(y*(y*x))} = C_0$, and so $x\ (y * (y * x)) \in I.$

Making use of Theorem 7.6.6, I is a commutative ideal the proof is complete.

Theorem 8.17 [10]

Given I is an ideal of a BCK-algebra X. Then I is positive implicative if and only if the quotient algebra $({}^{X}/_{I} ;* , C_0)$ is a positive implicative BCK-algebra.

Proof:

Suppose I is a positive implicative ideal of X. Denote $u = (x * y) * z$, then

$$\big((x * u) * y\big) * z = \big((x * y) * z\big) * u = 0 \in I$$

By Theorem 7.1.6, we have:

$$\big((x * u) * z\big) * (y * z) \in I,$$

That is,

$$\big((x * z) * (y * z)\big) * ((x * y) * z) \in I$$

Hence,

$$((C_x * C_z) * (C_y * C_z)) * ((C_x * C_y) * C_z) = C_{((x*z)*(y*z))*((x*y)*z)} = C_0.$$

It is clear that

$$C_x * C_y) * C_z) * ((C_x * C_z) * (C_y * C_z)) = C_0$$

Applying $BCI - 4$ for ${}^{X}/_{I}$ we obtain

$$(C_x * C_z) * (C_y * C_z) = (C_x * C_y) * C_z$$

This shows that $({}^{X}/_{I} ; * , C_0)$ is positive implicative.

Conversely, let $\left({}^{X}/_{I} ; * , C_0\right)$ be a positive implicative BCK-algebra. By Corollary 7.1.7, $\{C_0\}$ is a positive implicative ideal. If $(x * y) * z \in I$ then,

$(C_x * C_y) * C_z = C_{(x*y)*z} = C_0 \in\{C_0\}$. By means of Theorem 7.1.6, we have $(C_x * C_z) * (C_y * C_z) \in\{C_0\}$. This is equivalent to $(x * z) * (y * z) \in I$. Hence I is a positive implicative ideal. The proof is complete.

Theorem 8.18 [10]

Given an ideal I of a BCK-algebra X, I is implicative if and anly if the quotient algebra $({}^X/_I ; *, C_0)$ is an implicative BCK- algebra.

Proof:

This is immediate from Theorem 6.9 and Theorems 8.16, 8.17.

Theorem 8.19 [10]

If we are given a bounded BCK-algebra X with the greatest element 1 and an ideal I of X, then $({}^X/_I ; *, C_0)$ is also a bounded BCK-algebra with the greatest element C_1.

Proof. It is need only to prove that C_1 is the greatest elerment of ${}^X/_I$. For any $x \; in \; X$, $C_x * C_1 = C_{x*1} = C_0$, this means that C_1 is the greatest element of ${}^X/_I$. The proof is complete.

The inverse of the above theorem does not hold as shown by the following example:

Example 8.20 [10]

Let $X = \{0,1,2,3\}$ and the operation $*$ be defind by the table

*	0	1	2	3
0	0	0	0	0
1	1	0	0	0
2	2	1	0	1
3	3	3	3	0

Then $(X;*,0)$ is a BCK-algebra. The ordered relation is as the right figure. It is easy to verify that $\{0,1,2\}$ is an ideal of X and $X/_I = \{I,\{3\}\}$, this is a bounded BCK-algebra with the greatest element $\{3\}$.

Definition 8.20 [10]

Let $(X,*,0)$ and $(X`,*`,0`)$ be two BCK-algebras. A mapping $f:X \to X`$ is called a homomorphism from X into $X`$ if, for any $x,y \in X$,

$$f(x*y) = f(x)*f(y)$$

Example 8.21 [10]

Let $(X,*,0)$ be a BCK-algebra and I be an ideal of X. The homomorphism $f:X \to X/_I$ defined as $f(x) = C_x\ \forall x \in X$ is called the natural homomorphism.

Theorem 8.21 [10]

Given an ideal I of a BCK-algebra X, then I is maximal if and only if $(X/_I\ ;*,C_0)$ is simple.

Proof.

Let v be the natural homomorphism from X to $X/_I$. If I is a maximal ideal of X, we come to prove that $X/_I$ is simple. Suppose there is a proper non-zero ideal B of $X/_I$, then $v^{-1}(B)$ is a proper ideal of X and properly contains I, which contradicts to maximality of I. Conversely, suppose$(X/_I;*,C_0)$ is simple. If I is not maximal then there is a proper ideal A of X such that A properly contains I, and so $A/_I$ is a proper non-zero ideal of X/I by Theorem 8.15 (b). This contradicts to simplicity of $X/_I$. The proof is complete.

9. Conclusion

In this paper, we introduce the most prominent topics of BCK- algebras such as: subalgebras, bounded BCK- algebra, commutative BCK-algebra, positive-implicative BCK-algebra. In section 1, we introduced the previous studies about BCK- algebra. In addition, we study the properties of BCK-algebras. Moreover, we study the subalgebras and the different types of BCK-algebras such as the bounded BCK- algebra, the commutative BCK-algebra and the Positive-implicative BCK-algebra. Furthermore, we introduce the notion of ideals of BCK-algebras and study the different types of ideals. Finally, we present the concept of quotient algebras and its properties.

References

[1] H. A. S. Abujabal; M. A. Obaid; M. Aslam and Allah-Bakhsh Thaheem, On annihilators of BCK-algebras, Czechoslovak Mathematical Journal, Vol. 45 (1995), No. 4, 727–735.

[2] U. ACAR and Y. ÖZTÜRK, Maximal, irreducible and prime soft ideals of BCK/BCI-algebra, Hacettepe Journal of Mathematics and Statistics, Volume 44 (1) (2015), 1 – 13.

[3] J. Ahsan; E.Y. Deeba and A.B. Thaheea, MAXIMAL QUOTIENT BCK-ALGSBRAS, IC/88/272 INTERNAL REPORT (Limited distribution).

[4] ALI BANDERI AND HABIB HARIZAVI, DUAL ANNIHILATORS IN BOUNDED BCK-ALGEBRAS , Jordan Journal of Mathematics and Statistics (JJMS), 11(4), 2018, pp 325 -344.

[5] Atena Tahmasbpour Meikola, Eight Kinds of Graphs of BCK-algebras Based on Ideal and Dual Ideal, Neutrosophic Sets and Systems, Vol.33, 2020.

[6] R. A. Borzooei and O. Zahiri, Prime Ideals in BCI and BCK-Algebras, Annals of the University of Craiova, Mathematics and Computer Science Series Volume 39(2), 2012, Pages 266–276 ISSN: 1223-6934.

[7] Grzegorz Dymek Normal, pseudo-BCK-algebras، COMMENTATIONES MATHEMATICAE, Vol. 51, No. 1 (2011), 99-107.

[8] Hail Suwayhi. H; Abu baker; Ismail Mustafa Mohd, On m-Derivation of BCI-Algebras with Special Ideals in BCK-Algebras, International Journal of Scientific and Research Publications, Volume 9, Issue 3, March 2019 ISSN 2250-3153.

[9] Hee SikKim, J. Neggers and SunShinAhn, Pseudo-commutative algebras, *Preprints* 2018.

Available at https://www.preprints.org/manuscript/201810.0486/v1.

[10] Jie Meng and Young Bea Jun, BCK-ALGEBRAS, KYUNG MOON SA CO, 1994.

[11] S. Khosravi Shoar; R. A. Boezooei; R. Moradian and A. Radfar,

PC-LATTICES, A CLASS OF BOUNDED BCK-ALGEBRAS, Bulletin of the Section of Logic, Vol. 47/ (2018), pp. 33-44.

[16] Marek Patasinski, IDEALS IN BCK-ALGEBRAS WHICH ARE LOWER SEMILATTICES, Bulletin of the Section of Logic Volume 10/1 (1981), pp. 48–50 reedition 2009 [original edition, pp. 48–51].

[17] Mohammad Hamidia; Akbar Rezaeia and Arsham Borumand Saeidb, δ-relation on dual hyper K–algebras, Journal of Intelligent & Fuzzy Systems, 29 (2015), 1889–1900.

[18] A. Najafi and Borumand Saeid, Solvable BCK-Algebras, Çankaya University Journal of Science and Engineering, Vol. 11, No. 2 (2014) 19–28.

[19] Qiuna ZHANG; Cuilan MI; Xinchun WANG; Yuhuan CUI and Yongli ZHANG, Negative Implicative BCK-Algebras, Proceedings of Annual Conference of China Institute of Communications, 2010.

[20] R. TAYEBI KHORAMI; A. BORUMAND SAEID, A NEW IDEAL OF BCK-ALGEBRAS, Analele Universităţii Oradea Fasc. Matematica, Tom XXII (2015), Issue No. 1, 181–188.

[21] Young Bae Jun, SOME RESULTS ON IDEALS OF BCK-ALGEBRAS, Scientiae Mathematicae Japonicae Online, Vol. 4(2001), 411- 414, Received July 24, 2000.

Printed by Books on Demand GmbH, Norderstedt / Germany